The Generosity of Birds

Birds, like a dharma bell, flit through these poems. "The smoke has cleared. Listen—/the birds celebrate dawn./ Who am I to wail/for those I couldn't save;" It suggests only love for all of it, the withering and the ripening, will sustain us.

Part One focuses on a daughter's ambivalent and ever-evolving relationship with her mother. "Really this has nothing to do with my mother, /... Yet there she is, / a viridian flicker/on the periphery of my gaze, the vibrational key/"

Part Two ponders, can healing ancestral traumas expand our capacity for compassion? Traversing births, deaths, ordinary moments and life-changing transitions, these poems bear witness to the full continuum. "Somehow, I navigate/ calamities and the small miracle of breath. Soften my resistance to the rapture and rupture."

This collection interrogates "love" with its vast implications. It addresses our heritage of fear and miscommunication, exploring ways to free ourselves from beliefs and behaviour leading to environmental destruction. "Makes me hope we can heal backwards;" These poems look back to look forward.

THE
GENEROSITY
OF BIRDS

THE GENEROSITY OF BIRDS

Laura Jan Shore

Concrete Wolf Press
Louis Award Series

Concrete Wolf Chapbook Award Series
Poetry
ISBN 979-8-218-32154-3

Cover art by Oceana Pearl Piccone

Author photo by Sarah Wood

Design: Tonya Namura using Gentium Basic

Concrete Wolf
PO Box 2220
Newport, OR 97365-0163

http://ConcreteWolf.com
ConcreteWolfPress@gmail.com

Table of Contents

THE GENEROSITY OF BIRDS

PART ONE

This is the ferment I grow out of.
—Rainer Marie Rilke

Broken Needles, Lost Pins

I'm a jumbled basket of cut threads and spools.
Black linen head scarves and darned socks,

icy steppes, sputtering candles and mules,
I come from tweeds, grey-green as rocks,

peat fires and sheep huddled in the mist.
Pea-soup fogs and cinders blown down city streets.

The hold of a ship, a young seamstress kissed.
On the washing line, frozen sheets.

Oak and maple leaves patchwork the lawn.
A cotillion, my young Gran, dressed in chiffon.

Hand-tatted wedding veil, generations worn.
Felted slippers, a cloak of vintage twill.

Silver shank buttons and cashmere yarn.
Cross-stitch needle point, a granddaughter's skill.

I'm the warp and the weft, from each of them woven.
A hand-braided rug from the rags, I have chosen.

Inheritance

I.

To escape her parents'
squabble downstairs, she climbs
out her bedroom window, feet dangling,
heart tumbling.

Her wrist glances against ivory petals
as she leaps
into the arms of the dogwood,
          knees hugging the spindly trunk.

Her feet read each brittle
branch like braille. She lowers
herself to the final drop.

Across the frosty lawn, a solace
of purple in last year's leaves.
On her knees, she uncovers
the first crocuses, their yellow tongues,
an arc of joy. She sticks out
her own tongue to taste the pollen.

II.

Rewind all the way back,

to a plaid blanket on an English lawn,
one dimpled toe in her mouth, their voices
rising like crows behind her, the other foot
kicks at the bunching clouds.

Or forward, to the scent of sage

gathered in the Rocky Mountains,
fiery Indian Paintbrush flowers,
bees humming half-drunk in the autumn sun.
She balks when Mother says, *it's time to go home.*
Tears threaten, but her father's muscled arms hoist her
high onto his shoulders and her cheek
nestles into his silky black hair.

III.

No matter what menaces
her dreams, morning by morning,
the generosity of birds lifts her.
She opens her eyes to receive
her inheritance of sky.

Then there's the waiting at the hospital
with her mother, who now has dementia
and has forgotten why they've come.
She seeks refuge in a finger of light.
Scarlet leaves caught in an updraft.

After her father's funeral, she finds
herself, groceries on one hip, face to face
with a doe and her fawn,
ears twitching, their liquid eyes attentive
as she sings two wobbly choruses,
of *Don't Think Twice, It's Alright.*

What I Want To Know

It's the only four-letter word
no one utters in our house.

I dare ask her once,
face flushed with fever, head in her lap,

staring up at her peach pit chin,
smoke of her cigarette curling from her nose.

Her considered reply, *What do you think?*

I think not when she refuses to sign me up
for dance class, *too clumsy.*

Not, after school when we find her napping on the couch,
dirty laundry strewn from baskets; dinner unmade.

Not when she yanks at the tangles in my hair
and threatens to chop it all off.

Yes, when she feeds me pink aspirin
and mends my hem.

Yes, when we share an air mattress
and ride waves to the shore.

Yes, when we walk in the forest
and she teaches me the names of birds.

Not when she slaps me for wearing lipstick
and calls me a whore.

Not when she says, *there is only one good man
in this world and I've got him, so you're out of luck.*

Not when she invites my husband to move in
after we've separated.

Yes, when she throws a party to launch
my first book.

Yes, when she says, *If I had a voice like yours
I'd never stop singing.*

No, when she no longer remembers my name.

No, when she thinks I'm her older sister, the one she
hates.

No, when she tells my dying father to send me home.

Yes, when she lets me kiss her cheek
and begs me to come back soon.

High Stakes

To win our mother's approval, best
    to be a bit wild,
but not too happy.

We could sprint into the icy chop of sea
or scramble up our hemlock.
My baby brother balanced: *Look, at me!*

higher than the roof, arms outstretched,
a hawk catching a thermal, then snap, snap,
 a trail of broken branches, as he tumbled,
winded, but seemingly unharmed.

She'd drag out the story yearly
with the faded ornaments,
her own home-grown Icarus.

I never asked if that was why
he kept his scorched wings handy, trying
again and again to reach the sun.

 I was too busy prancing
around her moods, high-kicking
like a cheerleader, scooting away
from the back of her hairbrush.

No one spoke
    of what Dad came home to
after a full day's work; laundry
unfolded, dishes piled in the sink.

He whistled as he washed,
whisking around the kitchen,

fixing her martini, while she sat
at the table and smoked.

Be a horsey, Daddy, we'd clamour,
but just as he was on all fours,
two of us astride his back, Mum
would complain of the noise.

She brooded her sorrows
    like children—as if we
were the step-children.

No one ever explained
    we weren't to blame.

Joy was an affront to her—
    a pesky crow in the garden
she'd like to shoot.

Jeanne d'Arc of the Suburbs

Over flannel pyjamas, she yanks pants,
then the nylon waterproof pair, two sweaters
and the down vest. Plastic bags wrapped
around her socks, slipped into boots.
Tugs on her wool cap, ties the hooded jacket
beneath her chin. Muffler wound
across her nose and mouth, she stumbles out
into the changed world,
into the crystal vault, flakes that bite.

Veiled in grey, the sun is a moon
and the trees are sheeted ghosts.
She slips and clambers over drifts.
To the crunch and squeal of each footfall, she adds
her voice, *Sakura, Sakura,* her breath buoyant, soft
as cherry blossoms caught
in the wind, they flutter in her chest.

No trucks grind their gears. No yellow school buses,
not a single tire track mars the pillowy surface of the road.
Houses hunch under their bluish, white loads,
paths un-shovelled, doors closed.
She wears the weight of her layers like a mantle.
Left behind are the domestic squalls before the fire,
her mother's pinched lips, the puzzle pieces chewed by
    the dog,
her younger siblings clamouring to come along.

Her footprints fill and disappear
as she will, one day, swallowed up in white.
Though she shivers, this is not fear,
but a secret comfort that melts in her veins like sleep.

Taste of wool on her tongue, she daubs
at the drip of her nose, wriggling her iced toes,
her swaddled fingers stiff.
Lumbering on, she's a ten-year-old
self-anointed heroine,

ten blocks to the corner shop
for essential provisions: cocoa, popcorn, milk.

Bleep

after Edna St. Vincent Millay

My hands at her red leather steering wheel caused her
      distress,
she gripped her seat, surprised to find
herself the passenger, her middle-aged daughter not
      the kind
of driver she'd trust, though I negotiated the traffic
      with zest.
She was unusually chatty and I tried to stay abreast,
the thrust of her story about her father, through the
      muddle of her mind.
She mentioned something we'd always suspected, her
      words designed
to capture my attention, *alcoholic*, she called him, *he
      was possessed.*
She was stoic at his funeral as if tears would be treason,
and now, the diagnosis, calcium plaques in her brain.
Stop lights, children crossing, it was back-to-school
      season,
and I wanted to pull over, to really listen, in hopes she
      might explain,
her voice drifting, but someone cut me off and I never
      learned the reason.
Car horns bleeped out her words. A conversation, we
      never had again.

What My Father Needs

Outside the door, my mother
squawks like an adolescent magpie.

She's forgotten you're too weak now
to make her toast and she's forgotten how
to use the toaster.

When I open the door, she's startled.
Why are you here?

I step back and leave it to you, Dad, to explain yet again.

I can help, she insists. *We don't need her.
Send her home. I don't want her here.*

The chafe of this dynamic
is as old as I am—only now its stripped,
bare as a backside peeping through a hospital gown.

Your breath is laboured. I heft
you into the wheelchair and hook up your oxygen.
Still, she clamours.

I toss out the needles that flushed your port
and tidy the medicine bottles.

In a voice suddenly robust, you declare
what I've hungered to hear:
She's our daughter and I want her to stay.

You, who never dared to demand anything, tell her straight;

I'm dying.
If you want to help, I need you to be
—pleasant and present.

Maybe it's your tone, unequivocal, that penetrates
her faulty synapses, but she blinks, becomes—
almost my mother for those last few days.

A Little off the Top

after The Barbershop, *Edward Hopper, 1931,*
oil on canvas, Neuberger Museum, Purchase, NY

Enter from the lower left corner,
down the steps and into the shop
where the manicure girl waits,
thumbing her magazine.

We settle in, vintage voyeurs,
a long line of wheelchairs and walkers.

Follow the green curve of the ornate
brass railing, the white triangle of light,
sharp edge of window, a striped barber's pole.

The docent asks, *What does it make you feel?*

Lonely, offers my mother. It's the first
emotion I've heard from her in a decade.

Still, she adds. *Nothing is moving.*
And later, *The girl is sad. She misses
her boyfriend.*

Doris, gnarled hand
clutching a stuffed bear, says,
It feels cold.

Stanley, who I never knew could speak,
bobs his bald head and mumbles,
The man is hiding his face.

A Brief Awakening

Who'd have predicted
it'd be *Grieg's Piano Sonata in E minor*
that unscrambled your syllables?
Played often on Dad's favourite radio station,
but you used to switch it off and grumble,
We can't hear ourselves think!

The elderly pianist milks the chords,
her torso swaying.

Your white hair fluffs
around your ears like a new-born chick,
blush of skin beneath. Your elegant clothes
forgotten, you're slumped
in your hospital gown, the broken hip
that will never heal,
right leg propped in the wheelchair.
But your one good foot taps to the beat.

The pianist swoops up and down the octaves,
then slows, rippling the notes. The air becomes elastic.

Your long gaunt fingers rise up to conduct
in puppet-like jerks.
You've always been my impenetrable
mystery, your slammed bedroom door, that room
we were forbidden to enter, your secrets
hinted at, never divulged.
Now your language
is a jumbled code. It's been three years
since you remembered my name.

The room smells like damp wool and Clorox.
Veins river over your cheeks and nose. You'd hate

me to say this, but you look
like Granny, her final years, spent
silently stroking a stuffed teddy bear.

If I get like that, kill me quick, you begged.
Harangued my brother
to buy you lethal drugs,
just in case.
He refused, of course. Maybe that's why
he never visits.

In a flourish, the music crescendos,
sings sorrow, sorrow.
I've come to say, *Good-bye.*
By morning, I'll board the long flight back
home to Australia.

The final chords trail off.

And you are clapping, the only enthusiast
in an audience of empty eyes.
The musician strolls to your side, cradles
your ruined hands in hers,
looks deep into your grey wells.
Thank you for listening.

Suddenly,
you are fluent in your old decorum.
*Oh, thank you. And have you met
my eldest daughter, Laura?*

Regent Honey Eater

When I spot you, flitting
      amongst the eucalypts, sporting
yellow and black stripped
tail feathers, I feel

flattened, as I so often did in the presence
      of my mother, my wings outspread, a pin
through my thorax where she kept me
safe behind glass.

A birder, cultivator of gardens, she'd have pointed out
      your buttery-speckled breast, your nectar-sucking
beak
once ubiquitous, now rare, solitary and

like my mother, disoriented
      by clear cuts and suburbs. So clever
at mimicking others, you've forgotten
your own song. That broken song

my mother never sang,
      the one I was desperate to learn,
those missing notes to call in
a perfect mate.

Really this has nothing to do with my mother,
      gone for more than a decade.
Her ambition was to have no ambition.

To disappear. Yet there she is,
a viridian flicker
      on the periphery of my gaze, the vibrational key
as a bird call stiffens me to attention.

Bare-rooted Rose

after Li-Young Lee

I.

No thorns on the bouquet
I bring to my mother, rosebuds
still tight and white. My mouth
gapes, but no words escape as I watch her
dead-head the blooms.
Her sharp fingernails recall
the well-tended garden, her mind
has long forgotten.

II.

Withered blossoms, crimson-black
in the vase when death
first visits my parent's house.
Thick perfume clings
to the back of my three-year-old throat.
Their closed bedroom door.
My still-born brother. His name
never spoken.

III.

I learn to twirl. My own petals,
a ruffled pink petticoat.
If I spin out and out, will I melt
the frost behind my mother's hazel eyes?
My laughter is a thorn,
it pricks her.

IV.

A long-stemmed buttery rose
washes up onto the sand, tangled
in pungent bladderwrack.
A sprinkle of ash on the foam.
All the tender truths
I could never tell her.

V.

Her rose garden is leafless
now, all tortured angles
of shorn branch and thorn.
I stand bare-rooted on frozen ground,
then kneel to mound
fresh hay around each crown.

To Love

I.

Not once, did my mother say your name,
though we kids begged her to tell us.

What do you think? she'd ask,
fingernails tic, tic, ticking on the breakfast table.
For her, you might as well have been a curse.

Does it fall from skies above? I trilled
in the high school musical, yearning
to be dunked in the living waters of you.

Now your four letters have become common
as a copper penny, spent often, worn by many hands.

No, not a penny, you're more like a silk
parachute. Unpacked, you float me safely

back to dirt, but when you fail, I'm hung, dangling in a tree.

II.

The surprise of you—that first winter night with Anand—
        the lush timbre of his voice as he invoked your name

and everything became possible.

Ubiquitous as air, you're a verb
I declare again and again with my children.

Though we warned him, Anand persisted,
until he melted our mother's allergy to hugs.

I cradled her urn as we drove to the cemetery, womb
that forged me. How could she

not have noticed—you—were always there,

breathed in with ordinary intimacies—
a granddaughter's gaze, a just-ripe pear.

You circulate through us, spiritus as mist, always there
to be claimed—named or unnamed.

Reconstructing Mama

It's the thirteen-year-old version of you
in this pastel portrait, dated 1938,
your braid tied in a big blue bow
      (just months before it was lopped off
in favour of a perm).

Your eyes were already a wary
grey with the whiff of defiance
you showed years later by marrying a Jew.
In your sailor's blouse, you sat
stiff-backed, bored with this stillness,
the artist working swiftly to capture
your mouth's resignation. Like a skittish fox
freed from the trap of boarding school,
      (your gifts lost in the welter of dyslexia),
you dreaded the holiday's end.

Your picture hung in the hall
with the others, the third of four
siblings in sailor tops, a clever tribe
specialized in terrorizing the nannies
until they quit, leaving your mother in tears.

As I soften my stare, I wonder,
were you already determined
to become the mother you never had?

Not a pillowy Mama to snuggle up against,
no, you were "Mom", so cool with your Kent cigarettes
and diet pills, you encouraged cartwheels
in the living room, free-range-dinners
snatched from the fridge.

None of us, not my father who adored you,
nor any of your four children
could snap you out
of the melancholy that ripples
in soft strokes from this picture,
your face, faded and spotted now.

But what if
        it had been otherwise...

What if you'd been born
        as my granddaughter was
into a hot tub, a wise Hawaiian midwife presiding,
your father cradling your mother, both raising you
to the surface to marvel
        at your perfection?

I try not to explode
in mirth at the very idea of Grandad,
the Colonel, holding Granny in labour
because that could only have happened if...

she, too
had been a welcomed girl child and somewhere
in Ireland, her mother's
birth had been rejoiced over. In fact, we might
need to sponge out
the whole history of patriarchy.

Instead, I conjure a rebirth for myself:

in which you, my resolute Mama, rise up from the table,
free your feet from the stirrups, push away
the doctor with his forceps

and manipulations, trusting your instinct
to kneel on the floor and I
follow my urge to slip out in a gush, feet-first,

        so that the first gaze
you greet me with
and I mirror back to you,

is not one of disappointment,
        but triumph!

PART TWO

My looking ripens things and
they come toward me, to meet and be met.
—Rainer Marie Rilke

How to Love Without Being Uprooted

To prevent jelly-legs and snaggle-tongue,
    when your special someone comes near,

be as indiscriminate
    as the sun, radiate out

like an oak tree, roots and branches,
    in the effortless embrace of gravity.

Drop into the loud whoosh, whoosh,
    warm and viscous, the fluid world,

just behind your ribs, insistent song,
    attuned to your particular frequency.

Sequestered in those curved chambers,
    you will notice everyone's gaze

is a dance of mirrors, bouncing you
    back to you.

Practice curiosity. Every pore,
    a listening ear.

This is home, not an exile.
    Here you can be calm, but not soporific.

Don't let honeyed kisses divert you
    from the steady thrum of the hive.

Alone, yet together, drenched
    in appreciation for all of it.

As you giggle at the joke of otherness,

savour the heat of your palm on the sole
of your own upturned foot.

The Lie

The only significant lie she ever told
was tossed off into a crowd of strangers.
The air smoky, strobe lights pulsing,
her hips twitching, I watch her flirt
in imitation of the woman
she'll never become, flicking
her waist-length hair, a little daisy
painted on her right cheek.
It's Easter week, 1968, a crush
of bare-footed, blue-jeaned dancers
at the Electric Circus, but this man, dressed
in a three-piece suit, perspiration beading
across his broad forehead, is asking her to guess
how many BTUs of heat people are giving off.
That week she'd torn open the acceptance letter
from the college of her choice, had snuck out
to Greenwich Village in rampant expectancy,
to pose as already a sophomore, not wanting to admit
she's only 17. They groove to the Chambers Brothers live,
Time Has Come Today, the full 14 minute version,
and she allows herself a deeper gaze
into the tidal swirl of his grey-green eyes.
A firm voice alerts her—was it mine?
—*This is the father of your children.*—
A snort of laughter at the absurdity,
his face, a question she doesn't answer,
instead she darts to the Ladies, to straighten
her fishnet stockings, her flamingo pink
mini-skirt, oblivious to how young she really is,
but the old woman watching—myself, in fact,
50 years later—bites her lip.
She assumes he'll disappear, allured by another. But
he's waiting at the door, a grad student it turns out,
at the university where her father lectures,

he'd heard him speak and her throat tightens.
I watch them gyrate into almost morning.
There's no way to stop this.
He walks her to the train, treats her
to fried eggs at Grand Central,
scrawls her number
in green ink across his palm.

On Having Arrived at the
Age of Twenty-One

They also serve who only stand and wait.
            —John Milton

One white corner tucked
beneath her chin, she flicks

the sheet out straight and folds it in.
Her infant's in his crib, asleep at last.

She bends and folds, molasses movements
as the house exhales its emptiness,
six hours until her husband returns
to eat the meat she's marinating.
Transported safely back from war
on the zephyrs of her prayers,

in the attic, the fug of Vietnam.
Their letters and tapes mouldering.

Three years envisioning exactly this, and yet—

A surge sparks her nerves. The urge
to drop these sheets

and charge through the streets howling
like a wolf summoning her pack.

Instead, she sinks
on her haunches, sniffs

autumn breeze in the cotton weave on her lap.

Beneath her skin, a cloudburst.
Milk drips from her cow-heavy breasts.

Someone calls her name.
She leaps up, looks outside
but there's no one
at the door. No one.

Again, her name, a secret name,
soft then swells,

as if the bones of her skull
have become a bronze bell.
Heat flares up her spine.
Not one voice, but many.

In the silent intervals,
the mumble of rain.

Cathedral of sound, her name.
Palms buzzing, chin lifted, she stands

on the threshold, stands
gazing out.

The Interrupt

The interrupt has perfect timing,
taps your partner's shoulder
cuts into your dance.

The interrupt stalks you,
a long-distance runner
shadowing your every move.

It's an opportunity
you can't refuse.

A surprise party
just when you were heading for bed.

For Proust, it was the buttery
crumbs of madeleine.
For Dorothy, a whiff of poppies.

Weather is the supreme
interrupt. A frost that burns.

When the current is flowing,
the interrupt is a couple of beavers
chewing for one night.

A city's lights go dark—
and a new generation is conceived.

It can be a tick, the size of a pin-head
or a whale
surfacing in the moonlight.

The *you-you* of mourning doves
or the warning rattle of a snake.

You craft the form
of your life, the interrupt
provides definition.

Never romance the interrupt,
it sends him cold.

The interrupt takes you
in his arms, executes
a perfect spin. Leaves you breathless.

To a Lover, Forty-five Years Later

You never pretended you knew love
was more than a drunken grind.
The vicar never caught us

dancing naked in his garden.
And if two in bed was a lark,
why not three?

You kissed my lips until they stung
but never sucked my toes.
We hid the hedgehog

in the parlour and forgot to set it free.

I didn't ride behind you on your motorcycle
sparking the pavement all the way up the Khyber Pass
or follow you to the Isle of Wight.

We never promised
anything
and that was the whole point.

You failed to grip
my shoulders and shake me
to make me stay.

I failed to camouflage my parting pout.
We didn't stop writing love letters
after I flew home to marry

and if I hadn't burnt the lot, I might

have found you that drizzly April
when I returned to Cambridge
dragging husband and sons,

just as you'd predicted.
No, I never left a message
when I phoned and you were out.

I never imagine you without
your black leather jacket and pants,
your long greasy ponytail and boots.

Can't picture you
with a paunch or a stoop,
the tattered jumper of an aging Don.

You never were a gamble I lost,
more an ace held in reserve.
Don't call, just come.

Grandma Tess

Where've you been since we left you
on that sleet grey day in your plain pine box?
At the funeral, I heard your voice;
The suffering ends here.
Those words still sing in me
as I cuddle my own grandkids
or hang out laundry, your voice
in the flap of wet towels.

I see my small hands stretch up
to catch the sheets you cranked
through your wringer, stiff as cardboard.
Me, struggling to carry the bucket of pegs
in winter when the laundry froze and you beat
off the ice with a stick. Arthritic fingers stirring
buckwheat groats and poking the roast.

Standing tip-toe on a stool beside you,
the volcano of grains bubble and spurt.
And now, your scrutiny, as I soak beans
or air out my closets, remembering your house,
the smell of grandpa's cigars, mothballs, his lewd
jokes that never failed to make you squirm.

So teach me, Grandma, how to ease my pinched heart
for each overheard woe, the kind you brewed
into a cancer in your gut with a *tsk, tsk,*
toss of your head, braids coiled
around your ears like a crown,
flowered apron draped over your lumpy waist.

Wringing your chafed hands at tragedy
in Africa or next door or on your favourite TV soap,
wiping your tears with a dishrag as we listened

 to *La Boehme* on the radio, *I keep hoping
this time, she'll live.*

After 60 years of marriage
without a driver's licence,
you peeled off in a trail of clouds,
ignoring Grandpa's hysterics.

Death was your Chevrolet.

Phoenix, 1979

There's a baby blue electric blanket
at the drug store across the street.
He visits it daily, checking the price,
the same raised eyebrows and shrug, *$20?*
It becomes part of his routine, returning home
to phone his broker, his assets
all willed to his two middle-aged children.
He's proud of what he'll leave behind.

It's for my wife, he tells the store manager.
That woman of mine is feeling mighty low.
Taking pity on the old man, the blanket is sold
for half price. A bargain for his *best gal!*
He brings it to her with as much fanfare
as when he'd bought her the electric ice box in '36
or the wringer washing machine in '49.

Draped across the worn-out sofa,
she musters a grateful smile.
Please don't have pain, he whimpers, *think of me.*
By morning, he proclaims her better.
You were cold, is all.
In her dreams, she's being smothered
by fiery bedclothes.
I'll tell my cousin we can come to dinner.

They've had this conversation before.
Just try, dear, for my sake, dear.
You go, if you want to.
Not without you, dear. Put on your blue silk dress.
But, you know I'm poorly at night.
Don't give up, for my sake, don't give up.
I want to give up.
He tries out his old joke,

*After 60 years of marriage, we still don't agree.
I should sue for divorce!*

Revealed

Nothing much lovely about Grampa Lou,
not the reek of his cigar, the ash and crumbs
tumbling from his vest as he snatched us up
onto his lap, not his prickly moustache kisses.

He'd suck his false teeth at meals, slurp soup
and slam the table in a pique, upsetting the gravy.
Made Grandma blush and squirm
with his salacious puns and Mae West jokes

and who didn't wince at his tenor trills
while listening to Sunday night opera?

He pranced like a circus bear spouting Russian,
though he was only 12 when he'd arrived at Ellis Island.
Waving his cigar, he'd brag about the two jobs he'd worked
to pay for law school at night.

Weeping was a fine art for him and while Grandma lay
      dying
he wailed, *Mummy, don't leave me.*
The old aunts rolled their eyes and muttered,
About time she went somewhere on her own.

At the nursing home, the staff learnt to avoid
his flirtations and the occasional pinch.
By 96, still healthy, he'd had enough
and refused to eat.

Cocooned in white blankets, he was
a shrivelled balloon minus his bluster and puff.

Groaning in his sleep, wrestling with bedclothes,
with beckoning angels, he'd cry out, *No! No!*
raising his palm to ward them off.

His eyelids flickered then snapped open.
What time is it?
One p.m., Grampa.
Seeing me, recognition dawned.
He asked after my children, recalling ages and names,
then drifted off again only to wake and demand,
What time is it?

Once he sat straight up, grasping my hands in his icy ones.
He leaned his grizzled cheeks close.
Eyes, brimming like Russian lakes, revealed

the tender boy
he'd so skilfully concealed
beneath overcoats of bravado.
A luminous boy, we'd never met.

In the light of that naked gaze, he whispered,
You are beautiful!
spoken to me and to the reflection
of that boy beaming back.

The bare room glowed and everything—
all of it—was made lovely.

Into my lavender suitcase, I pack:

A tiger eye from my grandson, my swimsuit
        and red velvet coat, a faded letter on lined paper

my sister pencilled the summer she turned 6,
When you come home, I will give you a kiss,
and mailed to me at Girl Scout camp.

A fistful of poetry books, Rumi, Neruda, Sharon Olds,
all my underwear, my sneakers, my gut barometer…
my voice to say NO, to say YES,
my songs of resilience, my role as elder sister.

I carry armloads of psychic space, my heating pad,
toothbrush, and rescue remedy.

My job description: to stand beside her, more unshakable
than the oak in her yard, to face the doctors and to hack
our way through the welter of decisions.

I bring clarity.
My yoga mat. My presence.
My good black dress.

I tuck in my list of allies and leave room for unexpected
        angels.

I stuff in the mountains, the wind
and the whole of Australia.
My love affair with breath.

My knack at cobbling together ceremonies,
calling in a circle of women
to usher her on

or to welcome her back
if she has a miracle recovery.

My experience with triage.
My talent for witnessing
without being drowned.

  A surfboard to ride the guttural wail
   surprising my own mouth.

What I leave behind:

my clenched fist, my dancing shoes,
the static on the phone between us.

Women's Locker Room

after Marilyn Nelson

In the shower, it assaults me,
a chemical scent so acrid
it cuts through the miasma of chlorine
and sweat. I roll my eyes,
picturing one of the perfectly coifed
women in matching pink aerobic wear,
polishing her nails. High school cheerleaders
grown into *ladies who lunch.*

A visitor to this club, I'm a stranger now
in our hometown where my sister still lives.
Whispers swamp my memories—my skinny
adolescent shame, Lillian Wilson's knuckles
against my chest, the sound of my head
smashed into the locker for fumbling
a softball, the glower of girls
forced to pick me on their team.

Warm water pummels my shoulders.
I pause, letting it sluice down my spine
before I towel off, revising my shopping list,
the food I need
for my sister's kids, while she lies
in the hospital
with her yellowed eyes.

Stepping out, wringing my bathing suit
between my fists, I stare
at the bare back of the seated woman,
her cropped grey hair.

Reflected in the mirror, marks of the swim cap
rim her brow, but my gaze falls
on her one tanned, leathery breast
and the scar arrowing
across the left side of her ribs
which she makes no effort to conceal.

I'd seen my sister's chemo port,
but not this stitched up absence.
Like an Amazon Elder, she shoots me
a casual grin, then keeps on
painting her nails, a cayenne red.

Walking the Foreshore

I.

Foam scours the sand.
The pulsing lighthouse peers through
clouds like lumpy breasts
hugging the hills.

My sister's dead of cancer at 52 and I trod
the beach in Australia, burdened
by the glee
of children shouting in Chicago streets
more than forty years before.

I hear the grinding gears of a truck, we weren't
allowed to follow,
though more than once
my sister joined the throng
of leaping children in the fog of DDT,

and I was dispatched to fetch her,
my stubborn six-year-old sister,
stomping her feet all the way home.

II.

Splashing in and out of tide pools,
he abruptly turns and stares,
Did God make some men evil?
            Or do they just think they are?

Cheeks pink, blue eyes fierce
as only a five-year old's can be
            and I want the faith

to assure him, *They just think they are.*

Want to cast my lot
in Voltaire's Best of all possible...

and take Kierkegaard's leap,
shut my eyes to evidence blathered
            in the evening news, trust

the Buddha was right.
            The dharma wheel turns.
            As above, so below.

I gape at the cold blue and wonder.

III.

Clouds form a lavender mask this morning
with two fiery peep holes for eyes
and just above the horizon, a long slash
of a mouth streaming light.

I've never stared into the face
of evil, though I've read about it
and my imagination's good.

I'd so much prefer to discount it—
the aberration of a wounded madman
or a fool bereft of empathy
rather than cold, clear logic.

I search within my cells,
from fingernails to grey matter, but
my imagination fails.

Finally, I stop walking and lie face up
in the sand, my palms at rest on my sternum,
feeling my tumbling pulse.

Stop, and really listen
to the churn of the waves, their steady refrain;
        And now and now and now...

Parsing the Grind

for my sister

It's the ritual, the swoon.
Nose into the bag
        and already—my synapses fire.

This is how you return each morning to haunt me.
        Cold water slopped
into my Espresso pot-for-one, the gas flares,
my ears cocked for the bubble and spit.

Aromatic bliss—you craved and yet
        denied yourself off and on for decades,
not black, not milky, you were all herb tea and soy milk,
though you'd sneak the occasional cigarette.

We were hiking when you mentioned it, soreness
under your armpit, probably the kayaking you'd done
the day before and something
        on your breast, how you pulled open

your blouse to reveal a swelling, like a mosquito bite
that didn't itch. My hand hovered—electric
        with shock.

We dialled your doctor right there
amongst the scarlet maples and oaks. Her diagnosis
        came in less than a week.

Surgery and a year of saying yes
        to chemo cocktails
led to that night you woke from a coma, sallow-cheeked,
cancer-ridden, our family gathered close.

It was midnight and you wanted—
          coffee, the real stuff, not decaf,

carried on a tray to your bed. We lifted
your head to sniff. A blink, a sip,
          the chalice sanctified. Your final act

still percolates through me
these past dozen years, waking
          to cradle the weight
in both my palms, this blue china mug. Raised

to my lips, I pause long enough
          to drink in your last defiant grin,
the steam tickling my nostrils, froth
swirled around my tongue,

a single, savoured cup—bitter
          —topped with creamy fluff.

Father

His legs crumple
and I cushion his fall with my body,

the two of us sliding down the bathroom wall,

his bare legs skeletal, bluish bulbs
of reconstructed knees, stiff arthritic hands.

My modest father sits in the stink,
the colostomy bag split and spattered.

He's listing to the left. I stroke
the crepe paper skin of his once brawny arm.

I try not to gag, will my throat to relax.

Wordless, on the cold tiles, we
lean into each other.

He, who was everything—
vital and out of reach.

Inured to the stench, I suck in the rhythms
of his breath, the physics of death—

our stillness amidst the wreckage.

Silence

to my father

Outside, flying foxes squabble
over palm berries, their leather wings
flutter the fronds.

Overhead, an engine, a jet plane
fades into distance.

I want to ask you,

Is there anything more
now that your bones have sunk

deep into the sandy loam—
your restless legs gone still?

Salt breezes carry a whiff of night jasmine
with the tidal crackle and swish along the shore.
A possum scampers across the roof.

Silence is not absence of sound—

not tuning out,
more tuning in,

as each atom
in each cell

becomes
a listening ear.

Wasabi and The Crow

My darling,
        loving you was like romancing air, nothing
to butt up against
            except your charm.

Orbiting each other, the magnetic attraction,
            the white space between.

When I was pawing the ground, head bent
ready to charge,
            you were lithe as a matador, your silken cape
            chastened me.

How you always knew, even before I did—
You read my silences like braille.
            Stop yelling at me! said with a twinkle
            to my tight lips.

Some days, you played the bull
        and I would flee, pretending to be a wood nymph.
            Your winks and jokes tickled my nose.
When I sneezed, you quivered through me
            from scalp to soles.

        Like the wasabi and the crow,
        how it shook itself, feathers ruffling
after taking a beakful of green paste from the trash,
        eyeing us, it returned again and again, shuddering
each time like the first. How I felt when you nibbled my
    earlobe.

    Alone on the beach under an empty sky, it all comes
    clear.

You promised me vastness—
            and delivered,
even before your irrevocable exit.

Now that you are safely dead,
        I unlock my jaw—
                teach my voice to roar.

Too Beautiful

The shawl I wrap myself in
        when I want to speak with you
is a little rough against my cheek.
        Hand-dyed flowers and filigree rest
heavy on my shoulders,

smell like sandalwood musk
        and fiery winter nights.

You heard my admiration
        as we passed the shop window
and you took note. A Kashmiri woman's handiwork,
        satin-stitched on black merino wool,
swirl of rose, amber, and burgundy.

I'd forgotten all about it
when it arrived in the mail. You couldn't wait
for Christmas, presented it in July,
        your lopsided grin, dark eyes
wreathed with amber threads.

Below my grateful trills, you heard
hesitation. I couldn't quite explain
my heritage of privilege, the raiment of shame.

Gifts were your "language" of love while I
        was more dazzled by the poetry of your gaze—

Now sheltered
        in your big-heartedness, I cling
to that place in you
        where I was known.
This shawl, too beautiful to own,

and you—
already five years dead.

Waking in the Tropics After
Dreaming of Snow

Through the window, she-oak needles shiver
under the shimmying weight of a wagtail.

My eyelids are at half-mast, languor
after a broken night's slumber. Fruit bats in the papaya.

I feel chastised by the young magpie begging for grubs.

My body melts into my chair aware
of the Brush turkey in the banana tree demolishing fruit.

A caucus of crows dive and call.
From the tangle of mangroves—a python dangles.

What lingers is the scent of snow. A North American
       forest:
pine cones, blue jays, the dark velvet eyes of a doe.

Wistful as the crested pigeons cooing on my deck,
I'm missing my raucous laughs with you,

our nest under the quilts, the notes I could hit,
your baritone lilt,

how my tongue like a fish, muscled its way
along the shoals of your skin until—

silent as a heron—
you scoop me up and toss me deeper

back into the gullet
of my desires.

To My Mind

Your capricious meanderings even robbed me
on my wedding night.
Under my husband's tender gaze,
          you catapulted me forward
with plans for the week then ricocheted me
into memory, my stung cheek, my mother's slap
just as his lips tasted mine.

No more distractions, I am watching you.

Allow me to drift on the rising tide,
that song in my chest, the sprightly pulse,
then sink of satisfaction.
            Stay out, when I'm wrapped in his arms,
rain tickling the roof. Don't flash the dozen eggs
that crashed to the floor this morning
or the possible mouse in the closet.

I'm watching you—back off.

Trapped in the cave of your imaginings,
I've had enough airless fear, enough midnight shame.
Raised to worship at your altar, to believe
my diplomas proved your pedigree, I learn in fact;
I don't need you   minding me like an anxious parent.
I am wise to your propaganda, all the ways you kill
                the hours.

I close my lids and glare at you,
          yellow fog of ceaseless chatter
          hovering behind the bridge of my nose,
I stare until my eyeballs throb.
          Like the cat crouched by the mouse hole,
          I'm watching you.

To the School Principal Who Rescued Me

When you saw my head bowed
        to the steering wheel, palm on palm
            pinned to breastbone,

I felt not despair,
        though the adrenaline
            roared. I was not focused

on the traffic or how late
        I'd be for work. My boys
            refused to get out of the car.

It was a prayer to dissolve
        their defiance, my pulse
            still yoked to theirs,

 a pause to recollect
        the womb and the portends
            of the heart. I had wanted this.

Tilt of two moons orbiting, their chant
        looped behind my back. Their squeaky voices
            bounced me back to me.

In the cave of the sedan, I raised my imagined claws,
        conjured up a lioness,
            the perfect pitch of her growl.

When you tapped on the glass, red tie
        askew, frowning concern,
            shame scalded my cheeks.

But you smiled and the ruckus stilled. You clicked
open the door and firmly
ushered the boys out.

Detour

I'm tramping a well-marked trail and scarcely notice
the clotting sky. My mind, an amber river
where the dying bathe, the smoke of a funeral pyre.
My boy is wandering, somewhere in India.
It's been too long since I've had news.

The biting air rouses me. A sift of white.
On my tongue, thick snowflakes
taste like stone. Cling to my eyebrows and hair.
Birch trees shiver bare branches.
In a splash of red, a cardinal lands.

Dusk settles early. The path has vanished.
My boot prints disappear. Fingers fist
inside blue mittens. Gusts nudge my back.
I strain to hear the highway, stumble,
in ankle deep snow.

Tawny eyes grip mine. A fox, brash
against the white, sweep of russet tail.
With a balletic turn, he prances off.

Anchored, in my little clouds
of breath and tumbling heart, a hum
slithers up my spine. A towering
presence behind. A hand paws
my arm. I whip around. Not a hand—
a spruce bough, abloom with snow. Hello!

Reminds me of my boy's slender back,
the slope of his shoulders, arms still gangly,
head sprouted high above my own. My hands

rise as if to rest softly on his scapula, the heat
of his blood alive in my palms. I'm walking
behind him, walking
        behind him, always.

Home is heaven having a bath

Black cockatoos trembled the forest with their cries
this morning just before the storm. Now rain
pummels my iron roof, lashes the windows
and I glory in full water tanks
and the first hot bath of wet season.
As my toes hover over steamy water,
I always hear your childhood squeals of delight,
splash of your tugboat, your spontaneous song:
Home is heaven having a bath.
Bending my knees, I sink
into the soup of myself. Suds lap my chin.
The dulcet tones of cello on the radio, mellow
and deep, like your voice is now.
I'm remembering my young mother body,
bony with worry and you, too slippery
for my grasp.

My long-lost journals finally arrived,
five decades of scrawl only I can read.
I soak in how I've changed, how
I've stayed the same, naked
amid the lulling waves of my own voice
on the page. And I want to say forgive me,
my pounding doubts. It seems my terror
of losing you began long before you were born—
with blood-soaked sheets as the baby before you
seeped away. This is the stain that blinds me
to your perfect wholeness. Sometimes, my love is an ocean
that wants to swallow you back. Now I see the gulf
we've built between us protects you
from my sweeping tides.

Let me celebrate the towering tree of a man you've
    become.
A screen telescopes your life into my loungeroom.
We sing *twinkle, twinkle* with your baby daughter
and I'm dazzled by the liquidity of space and time.
So far apart and still the rain spills from the same sky.

Full Moon Eclipse

The sky is a silver net, moon caught
in the branches of a teak tree.
After the long wait, my boy chops wood,
loads it into the stove. I boil pots of water,
lug them to the birthing pool.

Cheeks etched with new wisps
of beard, face like Pan framed
in tawny tresses, he confides terror
in a stolen moment, the moon swollen
as a mother's breast.
I don't know how—to be a father.

Moans swell and fade. My daughter-in-law
paces naked, arm-in-arm with the midwife,
long hair damp with sweat.

Clouds scud across winter stars. I point up
to the winking eye veiled
by earth's shadow.
I'm empty and full.

Ready for the tub, I tip in
lavender and calendula buds.
She submerges into the perfumed water.
Climbing into the pool behind her, arms
wound below her breasts, my son nuzzles her neck.

The ragged moon has disappeared, the room
 illuminated by six candles.
Wind creaks the iron roof.
Flames gutter. Worry shadows
the midwife's face as she listens
to the flutter of the baby's heartbeat. Whispers

Get the oxygen, and I flash back
to the same heart rate drop for my son
after 20 hours of labour, my legs
trussed up in stirrups, the scent
of ether masking the doctor's panic
as forceps tore through.

The midwife's voice is calm and firm.
This bub wants out. Stand up, let gravity help.
      I breathe with her, my own pelvic floor softening.

I woke alone in the recovery room
not knowing if my child had been born alive
or later, if the boy they handed me
was truly mine, that eclipse of time
and skin contact, I'm still trying to retrieve.

Dawn brightens. She musters
fresh strength. Our voices soar, *Sweet surrender.*
Three strong pushes. The head crowns. Neck,
shoulders.
Into the water, the body slurps.

Baby, baby, she croons. They both grope
for slippery limbs. Instinctive parents raise him
to the surface like a trophy.
His mouth opens, gulps first air. The three of them,
Botticelli pink, nestled and naked in the dark pool.

I stand outside that warm, wet circle
like the far-flung moon, full-eyed
in reflection.

Joy

Out the screen door
into the musk
of night jasmine, I slip on
my gum boots,
pass the termite mounds
through lemon-scented eucalypts
to squat, wary of jumping jack ants
on the old wooden seat, overlooking
the moonlit mountain.
A spiral of flying foxes
squabble over palm berries.
The air is charged with the swell
of cicadas, a koala's guttural growl.
My soles are pressed to packed earth,
head bowed, alone under a kite of stars.
There's something elemental about damp
toilet paper, the act of balance
over the hole, slapping mosquitos,
picking a leech off an ankle,
smearing the blood.

Unrequited

The sea is a collector...
          —Marianne Moore

Lured by the throb and suck
of high tide, I'm immersed in purple
 and gold clouds smudging
     a brightening dome.

Summoned too, by rainbow lorikeets
who screech in code as they savage
the banksias and the winged
whistle of crested pigeons,
crunch of dried pods, the whorl of grasses
beckoning up and over the dunes.

There a dazzle of jade silk unfurls
to meet the edge of the sky
and my heart is hooked
like those sea bream agape
in the glitter of wind.

Oh, to plunge
into the foam, into the pummel
and swell, a body
seeking home, bereft
of fins and gills, limbs
loose as seaweed, salt-scoured skin.

A pelican, that prehistoric
angel, hovers as riptides swirl.
Another drowning last Saturday.
Brine spilt
from young grey lips.

I kneel
in the froth, beached.

Storms

Lightening ruptures the sky. Rain sutures
the soil. Windows distort runnels of wet
into prisms of dull light.

On the veranda, a half-drowned magpie
eyes me with an indignant ruffle of feathers.
Our road is a river now. Two pelicans glide by.

Banked at the verge, a tangle of branches.
Wading in, water laps my knees,
all swirl and swill.

Wind slithers
beneath my collar, billows my jacket.
I breathe the sea's dark breath,
salt crusting my tongue until
head flung back, face to the spitting sky,
I mimic the curlew's cry.

Don't give me shelter, tear open the roof
of my skull. Flood all my plans.
Toss me like flotsam
on the waves. I thrive in the tumult.

Storms remind me—
even alone, I'm not alone.

And if they ask how it happened

we'll talk about the mist,
the benediction of moisture saturating skin,
the cushion of clouds
between us and the valley.

It was the children who believed.

We came to worship the last remnant forest.
Eucalypts with raw, patchy trunks,
tall spines, leaves like tattered wings,
plumes of dun green air. Our nostrils flared.

Below us, derelict buildings.
After the drought, the fires, the floods.
Animal bones mound on street corners.
The persistent stench of death.

We climbed because the children dreamt

of trees. It began as a thought.
They huddled together in a little copse.
Their small faces stern with intent,
backs straight, bare toes pressed into topsoil.

Without a word, hands began to scoop.

They buried each other's feet.
They planted themselves, refusing to move
until roots, like little hairs
pulsed from their soles.

In the Rainforest, I Pray for Rain

Not even a spit, though the air dangles thick
with lemon-scented gum.
Headstrong clouds shroud the sun.

Shadows brittle my cheeks.
My thoughts splash into fissures where drought
spreads like leaf blight browning the forest.

Southerlies tease the canopy,
tremble umbrella ferns and trailing vines,
threaten to gust this promise out to sea.

Fogged with love and loss, I catch
a flash of iridescent feathers; yellow, green, red,
a shiver of light.

Rainbow lorikeets, upside down
in a grevillea, feed on fiery flowers.
Messy eaters, they pollinate the trees.

With scuffle and shriek, white cockatoos
rise as one. I lift my eyes to their tilt and glide,
fluid streaks skirting the veiled mountains.

Far below, my scuffed boots.
I'm tramping alone—awash in the ache
of boundless beauty.

A thundercrack peels back the leaden sky,
pours molten gold across ragged peaks, *kintsugi,*
like a mended Japanese vase.

In a swill of grey,
the mist, like smoke,
reseals itself.

Aria

The smoke has cleared. Listen—
the birds celebrate dawn.

Who am I to wail
for those I couldn't save;

the bees, koalas, ancient trees.
Surely, I could cough up one note

of hope out of this dry riverbed, herald
the green sprigs rising through ash.

Spit, spit—a curse on cotton farmers, coal
diggers, water stealers.

Calling lost tribes, the ululations in my throat
are visitations by ghosts without names.

When I say, "I", who might I mean? Not these bones
wrapped in thinning flesh, not this wrinkled face.

I wait for silence to enter me like stones
tossed into a well.

This song is not a rocket ship veering through space.
This song's a dilly bag filled with seeds and grasses.

It's a water hole harbouring new life.

Hush, hush—a blessing for tomorrow's children
as wind ripples the sand like waves.

Outcry While Waiting for the Moon

How to describe the peculiar
pungency
just before dusk

when the sea is docile
with only a fringe
of wavelets dipped in rose gold?

A slight winter chill, but no bite.

This much is true.
Two pink galahs ride
the feathery branches of a she-oak.

Children are bundled into jackets
and settled in sand beside small fires,
waiting for the full moon rise.

I try not to think
about glacial melt,
the capricious fury of floods.

From a distance, our neighbours
nod. A contagion.
Fear borne on the breath.

The teenaged boy from next door points
to the horizon. A flipper emerges,
waves at us.

We stare, united for a moment.
Tail flukes arch and slap the surface,
signalling what exactly?

Phones up for photos,
then one by one, eyes
drift down. The flicker of a screen.

Later, a jet plane hums,
the first I've seen
after months of cities in lock-down.

I'm surprised at my own voice, hoarse
after so little use, calling out to the steel wings—
Where are you going, where've you been?

Lemon-scented Gum

It was always you I sought for comfort,
an ark of green, gathering birds.

No ears, yet you knew the timpani of rain, no mouth,
but your answer to the wind; a fringe of buds.

I leant my back into your broad trunk,
rarely remembering to seek permission,

my toes trespassed amongst your roots, fountaining
deep into the soil.

Your sanction was the scent of citrus leaves.
I've no vocabulary to match.

At your outermost twigs, the weight
of birds tolled like a bell,

each cell tremoring all the way down
until even the taproot chimed.

Chasing elsewhere, I wore you
in my spine and in the thick pads

of my feet. Behind my shut eyes,
you stood unshakable.

But when I eventually returned, I found the gap
in the canopy. Your cracked stump and hewn chunks

of trunk strewn, each a hidden garden, host
to lichen and moss.

I sprawled face down hugging your girth
with all my little griefs.

What did I know of falling? You were one, now
you are many.

My cheek furrowed into your flaking bark. I listened
to small insects burrow and discovered—

>To wither is another way
>to flower.

Well Past Midnight

The house was quiet and the world was calm.
—Wallace Stevens

Her book topples from her hand,
lands on the floor, open.

Outside, centuries of stars, muted
by street lamps, a crescent moon.

Cicadas shrill, go still, then shrill.
Slap of waves pebble the shore.

Salt brine, jasmine. Sweat on her skin.
Heft of her pillowed head.

Conversations circle her fishbowl
mind. Her dead father's smile.

Sleep is the perfection
she seeks and yet—she dangles,

sheathed in sheets, cocooned
in the yes—of her breath.

Her pulsing hands, her restless feet.

Around her eyes, muscles twitch
as layer by layer, her masks slip.

Beneath her ribs, a sanctum.

Tenderness
warms her veins

dissolves
the house, the world, and she's

swirled, phosphorescent
on the turning tide.

Circumnavigating an Elliptical World

My nervous system wants nothing to change,
wants to freeze itself like a mastodon
in the tundra, but now—that's no guarantee.
My mind reminds me everything must
change and fast. Knowing is not enough.

Over the phone, I heard my mother's last whistle
of breath, the nurse saying, *She's almost,*
no, she's gone and I, on another continent, pictured
her crown of white fluff, her long bony fingers
at rest after years of aimless fiddling.

Death after death, those places of absence, landmarks
on the map of my skin and yet, I can still feel
my seven-year-old limbs, taut with terror
and glee. Spinning again.
Stripes of sun blazed across closed lids, sweaty
fingers gripping steel rail, knees
bent, sneakers slipping. But it was not enough.

Faster, they shouted. Running alongside, the big boys
pushed the merry-go-round. Schoolyard,
 a country of jostling and din. My skull flew
backwards, glimpse of sky, stomach in orbit.
Stop, I bellowed into the blare and thrum. Into deep space,
I flung myself, a habit
I returned to often, free fall before the crash.

My shoulder burned, prickle of freshly cut grass,
spread-eagled on my back, gravity suddenly real. Not like
    Atlas
bearing the world, I was flattened into it, still spinning.

"Around the World in 80 Days" was the only film
I'd ever seen, that colonialist chase through time zones
in strange vehicles: a balloon, a cart pulled by ostrich.
I was shocked at the red slash on my geography quiz.
I was certain the world was round, *wrong*, it's elliptical,
something I took on faith until 1972 when NASA
revealed the great blue marble.
Nothing's gonna change my world, I sang, still sing.

In a London pub, I watched the moon landing
on TV. Sipping cider, I cringed at my countrymen's hubris.
I preferred my moon mysterious, untrampled,
but it was not enough.
Now mining magnates will strip her of her final dignity.

When movies became apocalyptic, I preferred not to watch.
Somehow, I navigate
calamities and the small miracle of breath.
Soften my resistance to the rapture and rupture. My
    children
have borne children and will our world roll on?

Under marmalade skies, smoke drowns the sun.
Drought, floods, and that great pandemic pause; cities
emptied, highways silent. A collective sigh,
that *Stop*, I'd been breathless for—all my life,
and it's enough, just to lie here belly-to-soil, dig
a small hole, plant my face, sniff the dirt and listen.
The earth is ready—always—to receive us.

A Continuance

for My Granddaughter at 21

> *i am not a separate woman*
> *i am a continuance*
> —Joy Harjo

Once your ambition was to be a mermaid.
Now you dive and surface with ease.

Wet tendrils of hair accentuate the planes
of your cheekbones, mirroring my own.
      The tide wraps around us and tugs
at our limbs, like my mother straightening my skirt
with an appraising frown.

How many ways I've celebrated your declaration at age 5—
      (holding your artwork for all to see)
"I'm so proud of myself!"

I want to believe you will never
      hesitate to shine.
That you'll be the one to escape
the river of mothers branded by self-doubt.

You were the seven-year-old who shinnied
up Mulberry trees to collect leaves
to feed your pet silkworms.
      Still only seven when you learned

of your father's infidelity, you stood
by the creek, hands on your hips
      and confronted me.
"Didn't you teach him
not to lie? Didn't you teach him not to cheat?"

Over my kitchen table, your self-portrait hangs,
in a palette of browns—your blazing
          eyes, a wisp of a girl, white flowers etched
into your skin, recumbent on a gash
          of raw red earth.

At 21, like you are now, I was already a mother,
a wife, stunned by
          the gilt-frame around my life.

          Makes me hope we can heal backwards;
my Mum singing, instead of sighing,
my Granny, in pink leotard, learning ballet.

When I whispered into your unborn ears
and we painted your mother's belly with dolphins,
          my prayer for you
was to always feel   heard and held
          in a circle of women.

We tied red yarn from wrist-to-wrist
to connect us   with the cord soon to be cut,
          our voices, like stirred coals, flaring up.

Now standing at my stove, you are stirring spring
onions in a hot pan. We eat what we've made
together, silenced only by the sound of the surf.

Bellbirds

We're on the hunt,
windows wide,
driving in low gear,
ears cocked
for the high-pitched needles
as the valley unfolds below.

Stop, I shout,
and we ease onto the verge,
climb out into mountain air,
the piercing peals,
tink-tink, tink-tink
polyphonous chimes
speckle the hills.

Acoustic acupuncture,
these invisible sparks
prickle
across my synapses,
render me empty,
stupid
with joy.

Our Task

> *But our task is to die...*
> —Antonio Machado

I yield the heft of my bones, sniff
the gauze of darkness and sink.

This is my practice.

I imagine oblivion as turmeric yellow.
It tastes like blood.

Outside, the flutter of bat wings, an unhinged
gate, frogs moaning

like deranged cows, all blur
into a nameless din. A soporific bliss.

Sheltering inside my skull, I peer
at the poem inscribed there.

The line I can't quite read.
From the corner of lax lips, sorrow crusts.

I want to know the blank face
of my absence.

Untethered to the slump of clay, I want

to rise up
to where things happen,

but there's no time.

In the space between—am I
liquid as a chrysalis

or gaseous—
each pore an aperture
    streaming sun?

A Septuagenarian Reflects

Today, I toss
my judgements out
with coffee grinds and lemon rinds.

Contractions and reactions
my body knows so well
—must go.

I chop up the reasons why
my happiness has yet to bloom,
stir it into hot compost and let it steep.

Once I was a righteous woman,
my piety,
a mountainous grandeur.

I knew how to behave,
to bow to reality and make do,
ever-tightening my sphincter.

Today, I let my belly swell.
Melting my rictus jaw, I let loose
a roar—

Get-off-my-back!
to no one in particular,
a flick, karate kick, and strike.

Today, I drop onto my knees
and plant a yellow Tea rose,
a hybrid named, *Eternal Flame.*

I mulch to thwart the drought.
Layer leaf litter with leftover
ambitions and banana peels.

Eyes Open

> *I want to enter death*
> *with my eyes open...*
> —Claribel Alegria

I want to enter death with a full stomach
after a feast of pickled vegetables,
a mix of sour, bitter, and sweet.

I want to enter death resonant
with the names of God chanted
by children, sound looping out and out,

relentless as the song
of the male magpie in spring.

I want my last inhalation
redolent with salt breeze, coffee,
and apple peels browning in the sun.

My left hand smoothing the worn vellum covers
of Rilke. In my right, the comfortable
heft of a fountain pen drooling ink.

I want a caravan of colourful tents
tucked into a garden bower, where I lie
stretched out on silken pillows, my cheeks

brushed tenderly by grandchildren
with wands of pussy willows.

I want to know
in every desiccating cell of my being
that I belonged.

I want to enter death
like a lover who has been cheated
of nothing.

Acknowledgements

Thank you to the editors of the following journals in which these poems first appeared:

"Aria" and "If they ask how it happened": *Messages from Embers*, 2020

"Broken Needles, Lost Pins": *Panoply, A Literary Zine,* September 2023

"Full Moon Eclipse": *Brushstrokes* 2021

"Grandma Tess" and "Detour": *The Blue Nib*, 2021

"The Interrupt": won 1st prize in the Martha Richardson Poetry Prize, Ballarat Writer's Centre 2012; published in *Award Winning Australia Writing*; commended in the Aesthetica Prize and published in *Aesthetica Magazine, 2015.*

"Into my lavender suitcase, I pack": shortlisted for Fish Poetry Prize, Cork, Ireland; published *Fish Anthology 2015*

"Jeanne d'Arc of the Suburbs": *Panoply, A Literary Zine* January 2022, Issue 20

"Joy": *Not Very Quiet* Issue 5, September 2019

"The Lie": *Aesthetica Creative Writing Annual 2018*

"A Little off the Top": Highly Commended, Gregory O'Donoghue International Poetry Prize, Cork Ireland, April, 2016 *Southword Literary Journal*; republished in *Australian Poetry Anthology Vol.5 2016*

"Our Task": shortlisted 2015 Adrien Abbott Prize;
published in *Live Encounters Poetry and Writing 2021*

"Outcry While Waiting for the Moon" and "On Having
Arrived at the Age of Twenty-one": *Poetry of Encounter,
The Liquid Amber Prize Anthology*, Liquid Amber Press 2022

"Revealed": commended in the Tom Collins Poetry
Prize 2012; published in *Westerly*, 2013; republished in
Best Australian Poetry 2013

"Silence": first published *Poetrix*, May 2013; republished
in *Prayers from a Secular World*, Inkerman and Blunt
November 2015

"Storms": *Live Encounters Poetry and Writing 2021*

"To Love" and "Lemon-scented Gum": *Love, 2023
Australian Catholic University Prize for Poetry Anthology*,
October 2023

"Walking the Foreshore": shortlisted for 2015 ACU Prize
for Poetry; included in their publication *Peace, Tolerance,
& Understanding*, ACU Melbourne 2015

"Wasabi and the Crow": *Coal Hill Review*, Issue 29, Spring
2022

"Well Past Midnight": *Not Very Quiet*, Issue 8, 2021;
reprinted in *Elements*, 2023

"What I Want To Know": *Australian Poetry Journal*, V4.
Issue 1

"What My Father Needs" and "Too Beautiful": *Poetry of
Home Anthology*, Liquid Amber Press November 2023

"Women's Locker Room": *Resilience*, August 2021
Australian Catholic University Poetry Prize Anthology;
reprinted in *Aesthetica Creative Writing Annual 2022*

With gratitude: to my poetry buddies, Megan Welsh
and Ann Maioroff, for their expert feedback; to all
my teachers at the Pacific MFA program; and to Lana
Hechtman Ayers and Concrete Wolf Press for their
kind support in bringing this book to life. With deep
appreciation to my granddaughter, Oceana Pearl
Piccone, for all the ways she inspires me and for the
stunning cover art. Thanks also to my lovely neighbour,
Sarah Wood, for her photographic wizardry.

About the Author: Laura Jan Shore

Born in the UK, raised in the US, and now living in
Australia, Laura Jan Shore's poetry collections include
Breathworks (Dangerously Poetic Press, 2002), *Water over
Stone* (winner of IP Picks Best Poetry, Interactive Press,
2011), and *Afterglow* (Interactive Press, 2020). She's
also the author of the YA novel, *The Sacred Moon Tree,*
(Bradbury Press,1986) nominated for the Washington
Irving Children's Book Award. Her work has been
published in anthologies and literary journals on
four continents including: *The Griffith Review, Magma,
Southword,* and *The Best Australian Poems 2013*. In 2019,
she received her MFA in Poetry from Pacific University.

She won the 2012 Martha Richardson Poetry Prize,
2009 FAW John Shaw Nielson Award, and the 2006 CJ
Dennis Open Poetry Literary Award. Her readings have
included *Poetica* on Radio National, The Brett Whitely
reading series, the Hudson Valley Writers' Center in
NY, Perth Poetry Club, the Sydney Writers Festival,
Byron Bay Writers Festival, and Queensland Poetry
Festival. She has held residencies at New Pacific Studio,

New Zealand, KSP Writers Centre in Perth, Varuna in
Katoomba, and the NSW Poets on Wheels tour, 2003.
President of Dangerously Poetic Press, she has co-
edited 14 books and facilitated poetry readings since
2000. She has taught creative writing and poetry classes
and mentored writers since 1980.

"Reading a poem aloud entrains the heartbeats of
strangers and I love to be part of this communion," she
says. "Poetry brings solace and renewal in a chaotic
world. My dream is to find fresh ways to connect poets
and poetry lovers so they might nourish and sustain
each other in a global community."